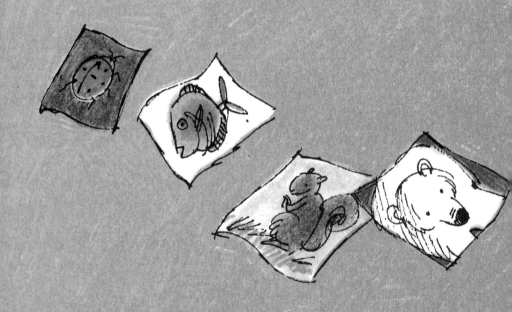

A SCIENCE I CAN READ BOOK

BENNY'S ANIMALS

AND HOW HE PUT THEM IN ORDER

by Millicent E. Selsam

pictures by Arnold Lobel

Harper & Row, Publishers
New York

For Susan Carr Hirschman

BENNY'S ANIMALS and How He Put Them in Order
Text copyright © 1966 by Millicent E. Selsam
Pictures copyright © 1966 by Arnold Lobel

Library of Congress Catalog Card Number: 66-10725

BENNY'S ANIMALS
AND HOW HE PUT THEM IN ORDER

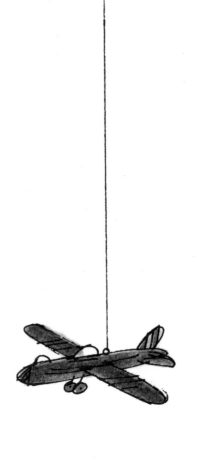

Benny was very neat.

He liked everything to be in its place.

He had a place for his large books.

He had a place for his small books.

And he had another place
for his middle-sized books.

8

If he had money,

he would put the pennies in one pile.

He would put the nickels

in another pile

and the dimes in another pile.

Once his mother said to his father,
"Do you think Benny is all right?
David's mother says David throws
his things on the floor.

And John's mother says
John never puts his things away.

But you know Benny.
He puts every little thing
in its place."

"Don't worry about him,"
said Benny's father.
"He is just a neat boy."

One day Benny went for a walk
on the beach.

He found many things.

He put them in a paper bag.

When he got home,

he took them out of the bag

and looked at them.

Some looked like this.

Some looked like this.

A few looked like this.

One looked like this.

And one looked like this.

Benny called to his mother.

"Mother," he said,

"do you know what these things are?"

"They are the shells of animals,"

said his mother.

"I did not know shells

came from animals," said Benny.

"Well, they do," said his mother.

"Some animals can take minerals

from the sea water to build shells.

This one is a clam.

And this one is a scallop.

I'm not sure what that round one is.

Let's look it up."

She gave Benny a book with pictures
of animals that live in the sea.
Benny looked at the pictures.

He found the round shell.

It was a moon snail.

He found the shell with points.

It was a starfish.

And he found the shell with legs.

It was a crab.

Benny went back to his room.

He put all the clams in one pile.

He put the scallops

in another pile.

He put the moon snails together.

He kept the starfish by itself.

He kept the crab by itself.

Then he made a sign:

ANIMALS FROM THE SEA.

Benny's friend John came to see him.

He looked at Benny's sign.

"Why do you call these things animals?"
he asked.

"Because they are animals," said Benny.

"That's not what I call an animal,"
said John.

"All right," said Benny,

"what is an animal?"

"A horse is an animal," said John.

"So is a zebra. And a cat.

And an elephant.

Animals all have heads and bodies

and four legs."

"How about birds?" said Benny.

"They only have two legs."

"Birds are birds, not animals,"

said John.

"How about fish?" asked Benny.

"They have no legs at all."

"Fish are fish," said John.

"I don't call them animals."

Then Benny asked,

"How about a butterfly?"

"That I know," said John.

"They are insects, not animals."

"But my mother said these clams

and snails were animals," said Benny.

"Hmmmm," said John.

"Let's go ask her again."

They went

to Benny's mother.

"Didn't you say my clams
and snails were animals?"
asked Benny.

"Yes," said Benny's mother.

"Any living thing
that is not a plant
is an animal."
"So what
is a living thing?"
asked Benny.

"Well," said his mother,

"our cat is a living thing.

But a rock is not.

Can you tell me

how a rock is different from a cat?"

"A rock doesn't eat," said Benny.

"A rock can't grow," said John.

"A rock can't breathe," said Benny.

"Very good," said Benny's mother.

"And a rock can't make
more rocks like itself.

Your clams and snails can do
all of this,
and they are not plants."

"So I guess they really are animals,"
said John.

"So my sign is right," said Benny.

That night Benny had dinner

with his mother, his father,

and his little brother, Eddie.

"Father," said Benny, "are we animals?"

"Well . . ." said his father.

"Well," said Benny, "we are alive.

We eat and breathe, don't we?

And we are not plants, right?"

"Right," said his father.

"So we are animals," said Benny.

"You are right," said his mother.

"But how are we different

from other animals?"

asked his father.

"We can talk, and they can't,"
said Benny.

"And Eddie is just like other animals because he can't talk either."

Just then Eddie said, "Da—da."

"See," said Benny's mother,

"Eddie is learning to talk.

So he is not just like other animals."

"But right now—" said Benny.

"Hush," said his mother.

"There's a big difference even now."

"Let's think of all the animals
we know," said Benny.

"Not right now," said Benny's mother.

"I have a better idea.

Why don't you look

at our old magazines?

You can cut out

all the animal pictures."

Benny got John to help him.

They both looked

for pictures of animals.

They cut out pictures of tigers

and butterflies and lions and snakes.

They cut out pictures of worms

and frogs and dogs.

They found pictures of monkeys

and whales and birds and fish

and lobsters and snails.

Soon they had a big pile of pictures.

"Let's put the ones

that look alike together," said Benny.

"All right," said John.

In one pile Benny put birds,
butterflies, and bats.
"That's because they all have wings,"
he said.

In another pile he put worms
and snakes.
"That's because they are long
and thin," said Benny.
In another pile he put
all animals with four feet.

"I'm putting all these
water animals together," said John.
And he showed Benny a pile of fish,

clams,

lobsters,

snails,

and jellyfish.

"Well, they don't look alike,"
said Benny.

"I know," said John,
"but they all live in the sea."
"That's no good," said Benny.
"I want them to look alike."

31

Benny's father came into the room.

"Well, how many animals did you find?"

he asked.

"Lots," said Benny. "Look.

We put the ones

that look alike together."

Benny's father looked at the piles.

"These are sort of mixed up," he said.

"Why did you put birds, butterflies,

and bats in one pile?"

"Because they all have wings,"

said Benny.

"But a bird has feathers,"

said Benny's father.

"And a bat has fur.

And a butterfly has scales.

I don't think they belong

together at all."

"How can we make sure?"

asked Benny.

"I'll take you to the museum,"

said Benny's father.

"Maybe somebody there

can help you out."

The next Saturday, Benny's father

took Benny and John to the museum.

"Can someone here

help us arrange these animals?"

Benny's father asked the man

at the door.

"Go to the fourth floor

and ask for Professor Wood,"

said the man.

When they found Professor Wood,

Benny spoke first.

"We have all these piles of animals,

and we are trying to put

the same kinds together.

Can you help us, please?"

35

"Let me see your pictures,"
said Professor Wood.

Benny showed the professor

the pile with worms and snakes.

"These do not go together,"

said Professor Wood.

"Why not?" asked Benny.

"They look alike."

"That is not

the important thing,"

said Professor Wood.

"We put animals together

if they have

the same *structure.*"

"Structure?" asked Benny.

"What is that?"

"It means the kinds of parts
an animal has and how the parts
are put together," said Professor Wood.

"The first question we ask is,
Does the animal have a backbone?
All animals either have a backbone
or they do not.
And this puts all the animals
into two big groups."

"Only two groups!" said Benny.

"That's just the beginning,"

said the professor.

"Why don't you go home

and put all your pictures

into two piles—a backbone pile

and a no-backbone pile.

Remember, any animal with bones

has a backbone.

Then come back here

and I'll have a look at them."

"All right," said Benny.

"Good," said John.

"Thank you," said Benny's father.

"Good-bye," said Professor Wood.

The next day Benny and John

looked at the pictures again.

"Do birds have backbones?" asked Benny.

"Sure. Didn't you ever eat a chicken?"

said John.

"Oh, that's right—bones," said Benny.

He put the bird picture

in the backbone pile.

"How about butterflies?" asked John.

"No backbone," said Benny.

"You can squeeze them."

"Jellyfish?" asked John.

"All soft as jelly," said Benny.

"Snakes?" asked John.

"Fred has a snake skeleton—

backbone pile," said Benny.

"Worms?" asked John.

"No backbone," said Benny.

"Fish?" asked John.

"Never mind. I know where it goes.

I see the backbone

every time I eat a fish."

"What about these shell animals?"

asked Benny.

"Those shells are on the outside,"

said John. "I once ate a clam.

It was all soft inside."

"Did you like it?" asked Benny.

"What's the difference?" said John.

"Put it in the no-backbone pile."

"I once ate a scallop," said Benny.

"It was soft inside too."

"All right—no-backbone pile,"

said John.

They put monkeys, tigers, lions,

elephants, deer, dogs, and cats

in the backbone pile.

"I am sure they all have bones,"

said Benny.

Benny and John did not know what to do
with some other animals.

They put them in pile number three.

Benny and John and Benny's father
went back to see Professor Wood.

The professor looked at the pictures.

"Very good. But what's the third pile?"
he asked.

"We did not know where
these animals belong," said Benny.
Professor Wood looked at the pictures
in the third pile.
"The crabs and starfish go
with the no-backbone group.
They are hard on the outside
but have no bones inside.

45

And the frogs go

with the backbone group," he said.

Then he picked up

the whole backbone pile.

"Remember what I said?

This is just the beginning.

Now we can divide this pile

into five different piles.

One pile will be fish.

Another pile will be amphibians—

animals that live in the water

when they are young

and on land when they are grown up.

Those are your frogs and toads.

A third pile will be reptiles—

your snakes, lizards, turtles,

and alligators.

Pile number four will be birds—

they all have feathers.

And the last pile will be mammals—

they have hair or fur.

When you go home,

divide your backbone pile

into these five groups:

fish, amphibians, reptiles,

birds, and mammals."

"What about the no-backbone pile?" asked Benny.

"You have enough to do now," said the professor.

"Can we come back later to work on the no-backbone pile?" asked John.

"Certainly," said Professor Wood.

"Then we'll know everything about how to arrange animals," said Benny.

"Everything?" said Professor Wood. "Not at all. You will know a little bit. Then you will have to find out how to divide each of the smaller piles into still smaller piles."

"Do you mean that after we divide
the backbone pile into five piles,
we can make still smaller piles?"
asked Benny.

"Oh, yes," said Professor Wood.

"There are about 17,000 kinds of fish.

There are thousands of amphibians,

reptiles, birds, and mammals.

Each group can be divided

into smaller groups."

"Well," said Benny, "this will take

longer than I thought."

"Oh, you can spend your whole life

on it," said the professor. "I have."

"Your whole life!" said Benny.

"Your whole life!" said John.

"Yes," said Professor Wood,

"but you don't have to do that.

All you want to know now

are the main groups.

Besides, knowing the separate groups
is not the most important thing."

"What is most important?" asked Benny.

"Well, you are learning how to separate
one group of animals from another,"
said the professor.

"That's right," said Benny.

"But I spend my time learning
how to bring them together,"
said Professor Wood.

"How can you bring them together?"
asked John.

"I mean that I find out

how animals are related to each other,"
said the professor.

"Do you mean animals have relatives?"
asked Benny. "Like cousins?"

"Yes," said Professor Wood, "but not
each animal. Each *kind* of animal
has relatives."

"Oh," said Benny.

"Do you have any cousins that look like you?" asked the professor.

"Yes," said Benny. "I have a cousin that looks a little like me."

"Do you know why?" asked the professor.

"No, I don't," said Benny.

"Don't you have the same grandfather?"
asked Professor Wood.

Benny thought.

"Only one of them is the same," he said.

"True," said the professor.

"Now can you tell me
why two different kinds of animals
might have structures that look alike?"

"Because they had the same grandfather?"
asked John.

"Sort of," said the professor,
"if by 'grandfather' you mean a kind of
animal that lived millions of years ago.
I'll tell you a true story.

About fifty million years ago

there was an animal

that lived in the forest.

It had a long body like a weasel.

It had a head like a fox.

It had a long tail.

It had sharp claws and teeth.

From that kind of animal

came lions, tigers, and leopards,

as well as small wildcats and housecats.

All these animals are relatives."

"I see," said Benny.

"You mean animals are relatives

if you can trace them back to—to—"

"A common ancestor," said the professor.

"Well," said John, "first we will put these animals in order."

"Then," said Benny, "we will become ancestor detectives."

"I never thought of that,"
said Professor Wood.

"That is really what I am."